VOLUME 8

HOW THE UNIVERSE WAS FORMED
Almatrinos & Urdires

FIRST EDITION

Carlos L. Partidas

quimicor2@gmail.com

Copyright © 2012 Carlos Partidas
N°   15873
Depósito Legal/Legal Deposit N°   MI2019000145

REGISTRO DE LA PROPIEDAD INTELECTUAL SAPI: N° 8074
DEL COMPENDIO LA QUÍMICA DE LAS ENFERMEDADES
REPÚBLICA BOLIVARIANA DE VENEZUELA, 07/05/2010
Derechos reservados/All rights reserved.

DEDICATORY

FOR ALL THE BEINGS THAT INHABIT THE UNIVERSE

CONTENTS

Chapter		Page
1	INTRODUCTION	1
2	THE INTEGRATING FORCES	8
3	INCREASE IN MASS	16
4	THE RELATIVISTIC MASS OF EINSTEIN	24
5	FORGIVE ME, EINSTEIN.	30
6	SPEED OF THE ALMATRINOS	36
7	EQUATION THAT FORMED THE UNIVERSE	41

RECOGNITION

To all the energy that animates all living beings
that inhabit the Earth

1

INTRODUCTION

Albert Einstein deduced that there is a relationship between the mass content m_0 of a particle and its amount of energy E, by equation $E=m_0C^2$. This mass is different from Isaac Newton's concept of mass, which should rather be called matter, because the mass m_0 to which Einstein referred. This is the mass of a particle at a subatomic scale, which is related to the motion of the particle but not to the force of gravity. And Newton's mass refers rather to the weight of very large bodies. In this way, Albert Einstein states, that particles by acquiring motion create their own mass, and that this gained mass m has to be considered in the following way:

$$V = \frac{m_0 C^3}{E}$$

E-1

Being m, the mass that the particle acquires when it is in motion, m_0 is the mass of the particle when it is still, V is the velocity of the moving particle and C is the speed of light.

And it is from this relativistic concept of Albert Einstein that we have deduced an equation that allows us to explain how the Universe was formed, and which is given by:

$$V = \frac{m_0 C^3}{E}$$

E-2

Being E, the energy that was generated inside the small bubble, that represented for that moment the incipient Universe, and U is the speed that the particle acquired, when this began to accelerate from its state of inactivity or stillness, or where still did not exist what today we call Universe. And this very small energy would be what Max Planck would call "action quantum", that is to say, the minimum amount of energy the system needed to propel itself. Or at that moment, the minimum energy was generated which the system was able to awaken by itself; and once the system has taken its impulse, nothing will be able to stop it.

But this equation $E = m_0 C^3 / U$; in spite of being very simple, gives an enormous sense to a complex but real case, because it shows us how the Universe was created. Because when the particle began to accelerate, the Energy E tended towards an infinite value. Therefore, within this small bubble, a relatively enormous amount of heat Q was generated, which made that microbubble burst, and it was what started the formidable energetic activity of the Universe. Because from this event, what emerges is only energy. And when we say that the energy was relative, what we mean is that that energy, however small it may have been, was too big to be supported by a system with those minimal dimensions. And this emerging energy, though equally very small, caused U, that is to say, the particle's velocity also increased towards an infinite value. That is, when E was very small, $U = m_0 C^3 / E$ ($U \to \infty$). And this case shows us, that $U = C^3$, that is to say that this particle managed to move at a speed equivalent to the cube of the speed of light. And this breaks the concept or notion that nothing can travel at a faster speed than light, or apparently violates Albert Einstein's Law of Relativity. But it turns out that this enormous speed, or C^3, can actually be obtained by extrapolation from experimental data. Only that Einstein did not want to see further, or when U/C was greater than 1 ($U/C>1$). Because Einstein did everything relative to the speed C of light, or everything he imagined and related when $U/C<1$ because he deduced that nothing could travel at a faster speed than light; because if this were so, in equation E-1, the mass m had to be necessarily imaginary.

But in this work, what we want to demonstrate, is that in reality, there are particles that can move at a faster speed than light. But because this kind of particle is the smallest that exists, we have had to identify it with another name; and because it has this different connotation, we have called it almatrino. Because it is different from the others. And because in addition to being the creative particles of the Universe, they are also the ones that gave origin to the spirits. So with this name almatrino, what we want is to connote an energetic trinity, because in reality humans and all living beings, we are made up of soul-urdires-consciousness, that is, soul-trine.

Of course, since these particles are the smallest, they have neither charge nor mass, because in order to create the Universe mass was not necessary; the only thing produced in the Universe is energy. And mass is formed when this same energy is able to condense through the intervention of other integrating or agglutinating forces. They are like glue. That is why these forces are also called from the English language glue, from which the term gluon is derived. And for this reason, or because our forces of union are different from the gluons, or in spite of fulfilling the same integrating function, to the forces that mediate to unite the almatrinos, we have had to call them urdires, because they are the forces that integrate, in a similar way to the process of weaving the threads to weave a fabric.

And the almatrinos with the urdires, formed another kind of threads of light so intense, that is what made us energetically stable, and to be totally independent Beings. And these energy entities can no longer disintegrate, because there is no longer an energy that has enough strength to break that union. But let's just imagine the whole marine population; a huge swarm of organisms, bees, ants, termites, plants, wild animals, etc., or the millions of sperm in the testicles of all the male animals, all formed by almatrines with the integrating force of the urdires! Or the more than seven billion equally different human beings; or the energetic entities that are part of the cosmic Beings; but also, those who have not yet incarnated, or need not participate in this system of life on Earth. And really, that on Earth, we can perceive life everywhere: be it as plants, algae, fungi, corals, spores, insects, mammalian animals, birds, humans, or whatever, and they are all really made up of another kind of energy, but in the form of agglutinated light. But at the same time, this is the only energy that is conscious of itself.

And it is evident that the Universe was formed from nothing. And something smaller than that point, which really does not exist. Because if the energy was very low, and the almatrinos did not have mass, we can assure that in the beginning, or in that place, nothing existed as mass. And only the least amount of energy that we can imagine, but that it was enough to disturb the smallest space that can fit in our mind. Logically, we will have to make an effort in our entelechy in order to imagine how small these dimensions are. But let us think only that within an electron we can accommodate about 10 thousand neutrinos, and within a neutrino we can introduce about 10 thousand almatrinos! Therefore, any physical material that exists can be transferred by these almatrinic particles, because for them, the electron or the nucleus of an atom would be too big.

In such a way, that the analysis we propose, we will do it using the existing laws of physics, or the ideas that are already rooted in the analytical mind of human beings, because in reality, that classical physics, or as it is, has only put us on that narrow path, or by highly complex calculation paths, and it is not precisely the idea of what we want to express in this book. Because with this analysis, what we really want is to give some logical sense, or an explanation to those phenomena that we can perceive or that happen to us. And the way of expressing in greater detail the emergence of life and the energy that forms the spirit, we have considered separately in the books "The Chemistry of Spirit" and "The Chemistry of Thought".

But Albert Einstein once said, referring to one of the greatest scientists in physics: "Forgive me Newton, but the deductions you made for large bodies are not fulfilled for very small particles". But then came one of those who most used the principles of quantum physics Stephen Hawking and said: "Forgive me Einstein, but the deductions you made do not apply to explain the phenomena of elementary particles". And apparently, we have arisen, saying with the almatrinos and the urdires: forgive me Einstein, but the almatrinos are the smallest particles that exist, but in addition, they can move with a greater speed than light. And, forgive me Hawking, because in a very simple way, this also contradicts the Big Bang theory.

Because as Albert Einstein himself deduced, because it was he who predicted that when energy travels at a great speed, it can do without mass to manifest its existence. Therefore, it is said that energy is equivalent to mass. In such a way, that the energy, although changing to mass and this one again to energy, both forms will always be real. And an energetic particle that manages to move with a greater speed than the light, of this high speed, is created, or the energy is transformed into an amount of relative mass.

But almatrinos can travel faster than light for several reasons: but let's say, among the first arguments, they can do it, because almatrinos are very small when compared to the size of any elementary particle. And, secondly, because nothing stops them in their trajectory; that is, it is impossible for almatrinos to collide with each other; and less against other particles, because there are no elementary particles smaller than almatrinos, so that they can be used as support for some rebound. In such a way that the nuclei of ordinary matter, as it was said, are too big for almatrines. Therefore, almatrines always travel in a straight line and without any obstacle that stands in the way. But in addition they can accelerate, and their acceleration allows them to reach a speed greater than light. And it was this enormous speed that caused the almatrines to produce all the mass that exists, and the mass that will exist in the Universe.

And obligatorily, that in the beginning, there must have existed a class of particles with a very low energy content. That is to say, that in time zero, or where there was nothing, we could not imagine that tiny Universe, but which in turn was very hot or very energetic, which is impossible, because it would force us to look for the origin of that heat. In such a way, that the Universe must have begun to form, by the movement of some particles with a low energy or with a very minimal acceleration. And from there, it was that the Universe began to form effectively, which now contains, or shows us its tangible and measurable physical part. But also, of these interactions, there remained an energetic portion and of matter that will continue being invisible, because they did not manage to integrate.

But the most complicated thing would be to know how long those particles remained in time zero; because if the particles are still, they do not interact, and the necessary forces are not generated to force them, or there is no what Max Planck called, quanta, or quantum of an action. In such a way that at the zero point, or before reaching an unstable situation to disturb the small space, a force must have arisen or originated; and this force was also imperceptible, the energy that motivated the beginning of the formation of the Universe. And yet that effect has not ceased nor will it be able to stop, and the Universe will not stop growing, since with the appearance of a new amount of heat Q, this will be greater and greater, but this in turn, will cause more mass to be formed, according to equation $m=m_0+Q/C^2$, or $Q=\Delta mC^2$; and to the same extent, or whenever a new mass m is formed, an amount of energy that has been generated in the form of heat will condense, but a new amount of heat Q will appear, which explains why the growth of the Universe is happening at an accelerated rate.

The almatrines and urdires were integrated and formed consciousness; that is, the conscious energy that animates any living being. For example, it is the fluidum or breath of life of the human being. So the almatrinos with the urdires are the forces that keep active that energetic flow of all the living Beings that can exist on Earth, but also the infinite classes of energetic entities that live in the Universe.

2

THE INTEGRATING FORCES

If we want to make a more real comparison of the integration of the almatrinos with the urdires, the closest we have found to understand how these intervening forces are, would be the gluons. But the discovery of the gluons, we are going to describe it in a light way, or only to make a necessary comparison, and to be able to have an idea of the immense quantity of forms of energy that they acquire, and how really small the almatrinos are.

And the probability of being able to grasp almatrinos seems to be too far away, because of how small they are, and there is nothing that can hold them, because there is nothing smaller than almatrinos to intercept them; or to leave a trace on the rebound. So, perhaps it will be impossible for almatrines to be identified.

But let's say, to compare it, that a gluon is what is known as a "vector boson", which means, that it has a value of spin one. Bosons are forms of energy, which have a value or swirl number, which is also called a spin, but its numerical value is an integer. For example, the Higgs boson has a zero spin value. The gluon is the binding force, which binds the quarks together to form denser subatomic particles called hadrons. They also intervene to hold the atomic nuclei together; and the gluons themselves can interact among themselves, or between the gluons of other quarks or the gluons of other nuclei, or by exchange between the same gluons. That is why we can say that the urdires, because they are more intense integrating forces than the gluons, they can interact to integrate the almatrinos, but also, integrate among them, and form other kinds of energies where we include the spirits. Which can perform independently, but form a kind of energy that is, or is conscious of itself.

But in the gluons, these interactions are so complex and varied that instead of the positive-negative electrical charge we normally know, the gluons interact through another kind of energy force called "color charges". And this kind of charge had to be defined in this way in order to explain or give meaning to the different kinds of interactions mentioned. In such a way that we can think that there has to exist another kind of interaction between the urdires. Or maybe the force of integration between the urdires was so intense that no more unions were achieved after the spirits were formed. Because there was no longer a sufficient energetic force, or one that would be able to continue integrating more almatrines. But it was so, that they were able to form the yotta-configurations energetic stable; but that in addition, they were able to differentiate between them. Because if you turn and look anywhere, you will see that there will be a Being different from you, or even if you try to look for one exactly like you, among the 7 billion human Beings, you will not find it.

But as we said, the comparison is necessary, so that it can serve as a relative or real guide, or help us to understand a little better, if necessary, what we mean about the almatrinos with the urdires, which integrate another kind of energy, which we call rather conscientia. And we call it conscientia, to characterize this energy that activates, because we will not know, if it is correct to call spirit, the energy that drives, for example, to an ant.

But following with the concept of the gluons, the color assigned to the electronic charge, if we can call it so, is a type of charge similar to the physical electric charge we know, for example between the negative and positive poles of a galvanic battery. But by means of this notion of the gluons, it is as if in the galvanic battery, we had three or more poles connected. And in this way, three charging properties have been identified in the gluons, to which three colors were assigned: red, green and blue.

Now, in these electronic charges known from a battery, each pole has its counterpart, so with the positive pole necessarily has to be the negative, so that electrons can flow from the negative pole, where there is excess charge, to the positive pole where there is obviously a charge deficiency. And this difference in charges is what generates an electric current. In the same way, between the gluons, with each color a series of interactions are generated; that is to say, a color with its anti-color. In other words, the red charge has its anti-red charge, etc. And each charge of color, is the one that gives place to the different forces of union, which are responsible for the interactions; for example, these are the forces that bind between the quarks to form hadrons and the different interactions that occur between the same gluons.

Therefore, we can imagine that from the combination between the different charges of urdires, an infinite quantity of possible combinations of energy between the almatrinos with the urdires and between the same urdires also arise. Because it is the only way to be able to explain to us why we are so many and so different. And perhaps we will not know that if instead of three as in the gluons, there are formed rather seven classes of charges, because equally we could call them rather frequencies, as in the musical notes. For these forms of combination and interaction between the seven charges create an enormity of energetic forms; that is to say, the varied and different musical pieces. And it is from these seven notes, or forms of charges, that the infinite forms of spirits and consciousness arise, with their forms of life: say a human, a dog, a whale, a cat, a cow, a bee, a clam, a coral, an ant, a flower, a plant, a worm, a bacterium, a cell, a sperm, an ovum, and so on. But if we dedicate ourselves to elaborate a complete list of these combinations, it would be something impossible to be able to complete.

But some of these combinations are those that are anchored to the mass, and that form the infinite combinations of organisms that we manage to identify, but that are necessary bodies, so that the almatrinos with their urdires can carry out that anchorage. And this anchorage of energy with matter has to take place first in the cells, because inside the cells, there is DNA and RNA, which are the only structures that self-replicate, and give physical form to all living entities that exist on Earth.

The quantum theory associated with the interactions of quarks, hadrons and gluons is known as Quantum Chromodynamics. So we will not know if it will be necessary to formulate a new theory in order to explain the infinite interactions that occur between the urdires and the almatrinos, and it must be greater than the three colors identified for the gluons. And to refer to these interactions through a theory, we could call it "Quantum Urdirodynamics". Or perhaps it is that these forms are affected by an effect called chirality, i.e. by a combination of left-handed and right-handed almatrines.

In such a way, that now we can think, that the interactions that took place at the beginning between the almatrinos, created the necessary conditions, so that these same kinds of integrating forces arose, and new particles were produced. And surely the different forms of mass; and from there emanated the infinite forms of intelligent energy which were projected with their own force, speed, and a necessary amount of energy, to form so many conscious forms, in addition to the spirits which spread throughout the Universe.

And additionally, that with the energetic forces integrating the gluons, two other classes of particles were formed; but they differed, and these are named rather as bosons and fermions. But a key distinction between these two families of particles is that fermions obey Pauli's exclusion principle, which establishes that there cannot be two identical fermions simultaneously at the same fundamental level, or that they are occupying a level with the same quantum number. That is to say, that they have approximately the same position, speed and direction of rotation. Because this corresponds to bosons. For example, there cannot be two fermions that have a twist or spin $+\frac{1}{2}$, because the sum of the twist would be 1, and that value corresponds to a boson. In this case to a photon. Likewise, two bosons that have the value 1 at the same quantum level would give another boson whose energy value is 2, and so on. In such a way that two fermions can occupy the same level or quantum number, but one of them, has to have a value of spin $+\frac{1}{2}$, while the other one has a value $-\frac{1}{2}$. And perhaps it is from here that the problem of chirality arises, and the discrepancies that are formed between the spirits, i.e. the setbacks. For example, twins usually spend time arguing with each other.

It means that almatrines cannot be bosons, because at the beginning, or when they were formed, they would have united with each other to integrate only one. In such a way that almatrinos are really fermions. And in contrast, bosons obey Bose-Einstein's statistical rule, and have no such restriction. Therefore, bosons can be integrated with each other, even if they are in identical fundamental states. But if the almatrines were bosons, they would have grouped together; or merged, and the Universe would not exist with its galaxies, stars and planets. That is, we would not exist either as bodies or as spirits, because there would not have been any integrating force, such as urdires. Which, evidently that they are bosons, because they manage to integrate the almatrines and to integrate among them, to form independent energetic units and integrity. For example, the gluons, are the forces that unite, in order to retain the energy, so that this energy is maintained forming the matter, or in the form of hadrons; and of these hadrons the quarks and the leptons (mainly the electrons, muons and tau) that between all (leptons and quarks) form all the matter that is contained in the Universe.

And just as the gluons arose, they must have emanated another kind of force that integrated the almatrines in order to form the spirits. And if these rotating movements of particles did not exist, which is what generates these integrating forces, of course atoms would disintegrate, or nothing would have formed as matter, and the Universe would only be luminous energy but without mass.

Because undoubtedly, for an energy field to exist, particles must be in constant motion. For example, the electromagnetic field is generated by electrons, only when they are in motion. And the movement of electrons is needed to generate electric current, which can travel through lines or through an electronic circuit, or as it was said, between the poles of a battery, but not without taking advantage of that current or movement, so that the electrons can do a job. Because if the electronic circuit is disconnected between its two poles, the electronic current will not flow and therefore the activity will be null in that circuit.

The spin of a particle was discovered for the first time in the electron; and it was due to the German physicist Ralph Kronig, who suggested at the beginning of 1925, that this spin was produced by the autorotation of the electron. But when Wolfgang Pauli found out about Kronig's idea, he criticized it, and pointed out that in that case, that hypothetical movement of the electron spinning in itself, would have to be faster than light, so that the spin would be fast enough to produce the necessary angular momentum. And that fact of supposing that a particle could travel at a speed greater than light effectively violated Albert Einstein's theory of relativity. But this according to Pauli. However, Kronig was right; since mathematically speaking, the effect of a tangential set is sumatory and relativistic, and that is why we say that two spins from ½ can be added together and obtain the value 1, which is the value corresponding to a boson.

But also this property of gyroplane disappears when the speed of light tends to infinity. And this value of velocity above light was mathematically eliminated when the electron spin value was replaced by a numerical value, equivalent to half the value of the quantum number. That is, without taking into account the tangential orientation in space. But it also follows from this consideration, that for fermions, this rotation could be of opposite sign: from right to left or from left to right, or with inverted directions, (+½ and -½) but this does not apply for bosons, because this number is an integer between two values divided by two: 0/2=0, 2/2=1, 4/2=2, 6/2=3, and so on.

But if we wanted to have a visual representation of these interactions, we owe these ideas to the American physicist Richard Phillips Feynman, who dedicated himself to drawing these interactions graphically, in order to explain the concept. So Feynman managed to make them understood, or imagine through a drawing, how it is that a particle collides with its antiparticle, to form for example a ray of light. Because from the collision of an electron with its antiparticle, that is, the positron, a ray of light emerges; and from this radiation, the quarks emanate; then the hadrons that make up the quarks, and with the quarks the nuclei are formed, and so on, as in a complex cascade of particles and events, which also generates the search for explanations, by means of mathematical formulations, to be able to model or systematize this barrage of physical phenomena. But now let's imagine the infinite interactions of the almatrinos with the urdires, which would be really huge for Feynman to draw them. But let's imagine that it was possible to reach a condition of relative energy in equilibrium, and where no more interactions could take place. Or at least with the same intensity as at the beginning, because the energy of the Universe was decreasing, in the same measure that the bubble that contains the Universe, was increasing in size.

3

INCREASE IN MASS

Mass is a definition used to get an idea of the amount of matter in a body. It is different from the weight of the body. When energy is collected or trapped by the cohesive forces we have described, this energy, in turn, will become another form of energy that we call mass, but that is not necessarily the weight. Because weight refers to the force exerted by gravity on the mass. However, gravity does not influence the content of the mass of the body. Hence a certain confusion arises; for mechanical physics, the mass of a body is a constant which is influenced by gravity. Whereas for the physics of relativity, gravity does not influence the mass of a particle, and this mass is a function of the movement of this particle with respect to the mass of the same particle, when it is stopped or at rest.

That is to say, when a particle is in motion, an additional amount of mass appears in it. And that is why Albert Einstein said to Isaac Newton: "Forgive me Newton". Because for Newton, the mass m is the constant that mediates between the force and the acceleration of the body (f=m.a). And maybe it is, because the movement of large bodies is very slow, when we compare it with the movement of particles. But really, that this Newton equation is not fulfilled for particles, because it would be impossible to measure their weight. But in addition, in the big bodies this mass, if it appears, the same one will be when the body is in movement. But it will be a very minimal amount of mass, because in addition, it will disappear when the body stops. Because it would be impossible to make a large body move at a speed close to that of light.

Whereas, through relativity, mass m is related to the idea of defining true mass as the value of the force between the acceleration experienced by a body when it is in motion. That is, for Einstein $E/m=C^2$; that is, mass is related to energy by means of a constant (C^2), while for Newton mass is the constant unit. But perhaps the transcendental thing about this fact is that this phenomenon was demonstrated experimentally. In such a way, that this was definitively clarified, thanks to the great and audacious relativistic ideas of Albert Einstein, who predicted that the energy would turn into mass and the mass in turn will turn into energy, when this mass gets to move at high speeds and vice versa. But at no time did Albert Einstein refer to the weight of bodies.

Let's take an example to imagine when energy becomes mass: in a cup of coffee, energy is trapped in the substance that forms the mass of the cup, but it is also trapped forming the plant of the coffee plant, from where the coffee beans emerged, which is nothing more than equally trapped energy. Then the energy was trapped in the beans in the form of caffeine, so the infusion or drink of coffee also contains water, where the energy was trapped forming hydrogen atoms, which in turn were trapped with oxygen atoms, and so on. And in this way, the energy went through a series of transforming stages, until it became different forms of mass, which were consolidated energetically. But then the mass was transformed into different forms of mass. And the force of gravity can act on mass forming weights, because if there is no gravity, of course there will be no weight, but lack of gravity cannot make the mass of bodies disappear.

And generally speaking, any solid body, whatever it may be, is really made up of particles, whose energy is condensed in the form of mass, because the only thing that is generated in the Universe is energy. And when an energetic force is applied to a single particle, in order to put it into motion, if this movement approaches the speed of light, this particle will create an additional mass, but that is relative to its inertial mass. And the Universe is in motion thanks to these energetic forces. But along with this movement of particles in the Universe, there will appear an amount of mass m, which will be relative or additional to the rest mass m_0 of that particle. In such a way that the real mass can appear only when the particle experiences a movement, and then it can dissipate towards other forms of energy, when the speed changes towards relatively greater or lesser values.

But if on this formed relative mass, other forces appear that brake them and integrate them, then the mass will remain condensed. And depending on the intensity of these forces, solid bodies will form or disintegrate. And in this way, an incredible dynamism is maintained, which forces an energetic activity and perennial movement of the Universe, between energy and mass. But also everything that exists in the Universe. And for the Universe to exist, everything that exists in the Universe must necessarily be in motion. And nothing can be immobile.

And as long as it keeps appearing, or being created from that energy, the relativistic mass and vice versa, we conclude that definitely the Universe is not going to stop growing. This should not worry us as human beings either, because this great activity has been going on for 13,800 million years and nothing is stopping it. And this time is relative to an instant of 1.45×10^{-5} years, if we compare it with the time that a human being of 80 years old can have lived on Earth. So we would still have much to do, because new galaxies will appear; and with them, new planets.

But one of the most immediate tasks, and that is what we want to achieve with this book, is to contribute to changing the absurd way of acting of some human beings. That is to say, to sensitize its degree of conscience, so that these are deserving of inhabiting those new spaces that will be created in our unstoppable Universe. In such a way, that the primordial thing would be to live with a certain order within this great chaos, because that is really what should configure the essence of the human Being. That is to say, what shapes the Universe as matter and energy in the form of spirit, which is equivalent to saying that it is formed by the almatrines with the indestructible energetic force that integrates the urdires. And this energetic set is what animates the existence of all living beings without exception. But this energy belongs to everyone, and does not concern human beings exclusively.

In such a way that the Universe itself is nothing more than a physicochemical system, whose growth cannot be slowed down, unless all the immense energy generated is solidified in the form of mass, which is going to be equally impossible. And only that we will be able to ride on the created bodies, or tremble freely on them, because we can really move much faster than these bodies do in space.

But if 50% of the energy generated were to become mass, this would have to happen within 175 billion years, that is, when $E/m=1$. But since the energy is most efficiently contained in the form of mass, so far the amount of energy that has become mass represents 4%. But perhaps, that this amount is really a point of equilibrium between free energy, and the greater amount of energy that has been stored or trapped in the form of matter. And perhaps that 50% value cannot be reached, because a new amount of energy will always appear, which really comes only from movement. And with this force, the new space is also being created. And in order to be able to cross that new and immense space created, the only way to achieve it is that we can travel at a speed equivalent to the cube of the speed of light. If it is that for that moment, we continue taking that value of the speed of light as a reference; because we don't know if another way will appear to reference our concepts, ignoring the current physics.

And we begin to count from there the phenomenon of time, which is only useful, to have an idea of the before and now. And it would be better to visualize it as an eternal moment, because what has already happened cannot happen again, at least in the same way, but events will continue to appear constantly, from the moment when the Universe was only a very small bubble.

The Universe is now, but it will still be very large for us as physical bodies, but perhaps small if we can move with the speed to which almatrines move. And the only way to cross the immensity of the Universe, is that we can move with great speed. Because for example, if we want to reach the center of our Milky Way as spirits made by almatrines, it would take us approximately 9 seconds to make that journey. A distance that it would take light to make that journey, about 25,000 years. And just to get an idea, within 175 billion years, the Universe will have reached a size equivalent to 12 times its current size.

As for light, the photon phenomenon, was proposed by Albert Einstein, who was able to predict brilliantly that in reality light does not travel in wave form, but as packages of particles, which Einstein called photons. Whose term is derived from the photoelectric phenomenon. And the form and variety of these frequencies, is what induces us to think, that the urdires could adopt different energetic forms, reason why instead of colors, as in the gluons perhaps we could call it rather tonalities. Because it turns out that this theory to explain the photoelectric phenomenon was also demonstrated experimentally, by the American physicist Robert Andrews Millikan. And without going into details, because here we are only interested in the physical phenomenon, Einstein deduced that the energy of a single photon is given by $E=h\upsilon$, where υ is the frequency of incident light and h is the Planck constant. Or that $h\upsilon_0=E_0$ in the fundamental level or where the kinetic energy is minimal, that is to say there is practically no movement; and therefore also the frequency υ_0 is the minimum frequency. And applying the concept to the photoelectric effect, Einstein wrote, that $h\upsilon=E_0+K_{max}$ Where $K_{máx}$ represents the maximum kinetic energy that the electron can have, and that is sufficient to release another electron to the photoelectric material. And when υ is less than υ_0, the photons will continue to be individual, no matter how much they are, as Millikan proved. This means that the intensity of the luminous radiation is not important, because the photons will have enough energy to expel the photoelectrons. But it is the force that integrates the material that does not allow it to lose its electrons. Because this amount of E_0 energy is characteristic of the substance, and is said to be a property called the working function of the substance.

In such a way, that making a similarity of the energetic force between the urdires and the almatrinos, this energy that was formed, is also individual, and of course there is no longer an energetic force in the Universe, that is capable of breaking the integrating force of the urdires with the almatrinos, or that is sufficient to overcome the function of the work of the spirits to disintegrate them.

On the other hand, we have shown from Einstein's equation that all kinds of matter are absolutely real. But Einstein assumed the opposite. Since if Einstein said that energy necessarily has to be real, we must assume, because matter is actually the energy that has condensed, or from which mass was formed, of course, that we can reason, that all forms of matter have to be equally real. And everything that exists in the Universe is real. And we cannot say that the Universe is a hologram, or that its matter arose in some way from the antimatter.

However, following the passage to these particles, if after a while, these particles of elementary matter with excess electrical charge, and therefore negative, is achieved with its opposite, i.e. with another identical elementary particle but has a charge deficiency, or positive, which we identify as antimatter, only as a way to differentiate them. But when these two particles meet, they will mutually annihilate each other. And what will remain after that collision is effectively a luminous radiation. That is to say, from matter and antimatter, energy will emerge again.

But perhaps the most interesting thing is that from that same luminous energy, can emanate again the matter and antimatter, which are events that will no longer be able to stop. And if matter was formed by a force which is able to hold it together, then of course the energy can be trapped, until other forces arise which are sufficient to disintegrate it again. While in other cases, the integrating forces may be so weak that matter will disintegrate by itself or spontaneously, or by the incidence of just one ray of visible light, such as the photoelectric phenomenon, phosphorescence and fluorescence. And the whole phenomenon fits into the term of luminescence. And as for living beings, the process is known as bioluminescence, when light is converted into images, and sonoluminescence, when light is converted into sound, by molecules that explode and form again as if they were small bubbles.

But among other observations, all the antineutrinos (or neutrinos with positive charge) observed until now, have chirality in the turn, as Kronig observed it, and that the direction of this turn, is of right hand. In other words, its direction of rotation is from left to right. It is like imagining a whirlpool of terrestrial wind, whose vortex of turn is from left to right. While electronic neutrinos (or neutrinos with excess negative charge) are left-handed. Or that they have a turning helix that coils to the opposite side; or that their turning vortices are from right to left. And this is an extremely important observation in order to understand the property or behavior of matter; and therefore, the character of ourselves as bodies and as energy.

And this again, forces us to suppose, that the quantity of electronic neutrinos, that is, those neutrinos with excess of negative charge, became more abundant, with respect to the quantity of antineutrinos, because the forces that induce spontaneous disintegration became less and less, or was no longer sufficient to promote the formation of more antineutrinos. Because the most logical thing would be that the quantity of matter and antimatter had remained invariable, from the very moment the Universe began to form. That is to say, if the rays of light had gone out in the same direction, of course the quantity of neutrinos would have been completely annihilated with an equal quantity of antineutrinos. Or as it was said, almatrinos have to be fermions but not bosons. In other words, if all neutrinos with antineutrinos had been annihilated, matter would not exist; and our Universe would be like an inhospitable and radioactive desert filled only with light, or it would be only energy without any kind of matter. That is, there would not exist a form of energy trapped or condensed by the electronic force of atoms by means of gluons. And it was only because of that apparent cosmic anomaly that we actually have more matter than antimatter today. And that difference of energetic force, or charges, and annihilation, the emergence of radiation etc., is what keeps the Universe in constant motion, but also, thanks to it, is that we can say that we exist. Because if we were bosons but not fermions, then the almatrines would have merged into one. But also, there would be no stars, galaxies, planets, photons, molecules, DNA, black holes, and so on. That is to say, nothing related to matter would exist.

THE RELATIVISTIC MASS OF EINSTEIN

And this way the little Universe awoke from its lethargy or stillness. And the minimum amount of mass was formed which remained as m_0. And when the value of the minimum energy E was equaled with the value of the minimum mass, that is, when m_0/E in the equation $U = m_0 C^3/E$, it became equal to one, at that precise moment the speed of the almatrinos became equal to the cube of the speed of light. $U=C^3$. And the energy E was made equal to m_0 ($E=m_0$). And at the same time, that velocity created the mass m, from the energy E ($E=m$), which became relative to the mass m_0. And the mass was converted back into energy. Then, or immediately, the integrating or agglutination forces were created. And once this point was reached, in that minimum space, the necessary conditions were given and achieved, which disturbed that small system, which until that moment was immobile, and from there the beginning of the formation of the Universe was given, some 13,800 million years ago.

And just as Wolfgang Ernst Pauli did when he proposed neutrinos, we dared to call those particles that motivated the gestation of the Universe, as the truly elemental among the most elemental ones. That is to say, the almatrines. And from these, or due to that acceleration of the almatrinos from a value of zero velocity, the energy E became infinite; and that enormous energy arisen, with respect to that small bubble, was what created the movement, and the mass of Albert Einstein was formed. and again it was the same energy that formed other particles, such as neutrinos. And then followed the forces that unite or agglutinate; that is, the urdires, photons, bosons, fermions and gluons; and with them, the hadrons; and with the hadrons, the quarks, and with the integrating force of the photons, the electrons that formed the family of the leptons, and between the quarks and the leptons, everything that we can physically see in the Universe was formed. And with the almatrinos and the urdires, came infinite tonalities or different kinds of conscious energy, and among them the spirits. And everything that exists and what we see, but also what we do not see.

And they would be free, only the other almatrines that did not manage to integrate, because the energy needed to unite them faded away. But almatrines will continue to be the smallest particles that exist, and those that can be formed will be in greater proportion, or will accumulate to form part, perhaps of matter and dark energy that fills all the space that went, and is forming. Or what is now constituting the whole Universe. And we can say as for the Universe, that the expansive wave is continuously moving away towards the periphery of an enormous sphere, whose radius is increasingly greater.

But Albert Einstein was right, when he established, that if a particle moves at a speed, at least close to that of light, that particle will create a quantity of relative mass, from its inertial or resting mass. Because the relationship between energy and the square of the speed of light is precisely mass ($E/C^2=m$). According to what Albert Einstein correctly predicted.

So the very high speed of the almatrines ($UE/C^3=m$) greater than the speed of light, made the mass m appear, which would be a smaller mass than Einstein's mass, because instead of C^2 found in Einstein's equation, in the new deduced equation, appears in the denominator the speed of light C, elevated to the cube (C^3). But although this mass is smaller than Einstein's mass, it also involves the speed of the particle U, but also, this new mass, although very small, has a positive sign, so like energy, this mass is real, but can never be imaginary. And this really makes more sense than Albert Einstein originally raised.

And this minimum mass, became again different forms of energy, including the energy that was generated in the form of heat, and thus began to heat the Universe. Only that the initial space was very small, so relatively, that the interactions were very intense, when the Universe was still reaching the size of a traveling globe. And again the energy became mass, until all this caused a violent and unstable system to form, which was self-feeding like a boomerang and at the same time as saying, forming and disintegrating by itself through a process that can no longer be stopped. For he himself forms the mass and the energy which keeps the system perennially energized on its own. And we already know that if there is no movement between the particles, of course there will be no generation of electronic charges either; and it would be the only way, that the system could be kept inactive. In such a way that the process has to be necessarily motivated or activated by a constant movement. And it is going to be extremely difficult for the Universe to shut down.

But the only way to appease the intensity of these magnificent events is for the enormous amount of the different kinds of energy generated to solidify or condense into mass, and in this way the energy can be kept confined or united by the energetic forces formed by bosons, gluons, hadrons, quarks, leptons, urdires, photons, etc.

And then arose the electromagnetic forces that hold atoms together; as well as the electron rotating around a nucleus, from there the molecules, and with them, the energy in the form of matter that becomes visible and malleable to transform them into other equally infinite forms of substances. But this is only a combination of mass with mass, and in order for one to form, another has to disintegrate, and in that exchange, only one transfer of energy intervenes, although not necessarily of mass. And these new masses are nothing else, they are the same energy that emanated, but is now solidified.

From that exchange or interaction, DNA will also arise, from these cells, and from these bodies, which serve as enclosures to be occupied by the energy that was consolidated in the form of consciousness and as spirits. Although it would be very difficult to know if the other kinds of energy that form the living bodies, are not conscious of themselves, because I personally have managed to "speak" with a swallow and a hummingbird. Or who hasn't been able to communicate with their dog or cat, for example?

But hopefully the planet will not be destroyed before other brilliant minds of man can make that great leap to detect almatrinos. And it seems that time will not suffice for it, because other minds, subhuman or humanoid, live clinging to the ambition of wanting to destroy and dominate others on the planet, (rights against left-handers) and piercing the Earth to destroy it and extract its own body, arguing that they are resources that belong only to them, as if the Earth corresponded only to a group by a decree or preferential and divine signaling.

Selfishness has taken over some human minds, which only give value to that which is part of the material, to try to turn it into money in any way. And they even want to buy the analytical capacity of others; as long as these retarded individuals, with their money, can take economic advantage of what others have created, or that has a value that can be turned into money.

In that sense, to live in such a physical world would in fact be an absurd act. And when each and every one becomes aware of their origin, or as almatrines integrated by the indestructible energetic force of the urdires, and with thought, only in that way achieved, humanity and the new humanity can change. And those who by some custom, or economic power insist on bending others for no reason, but with the clear intention of the economic, will have to be taken to those primitive or less evolved planets, so that from there, as matter and antimatter annihilate one another, a transfigured energy can emerge from them, that can be more useful, or that do not persist in continuing to damage the harmonious coexistence of the great Universe. And in a general way, not to violate the right to live that absolutely all the other beings that exist have, and those that will occupy in their moment, the Earth as their dwelling or home only temporarily.

All forms of energetic and physical life are made by almatrines and the energy of urdires, so absolutely everyone has the same right to form physical bodies on Earth, with their great variety of forms and their different biological processes and purposes. But unfortunately, this energy catastrophe happened, even though other beings arrived millions of years before us, who just arrived 200,000 years ago; but in less than 200 years, we have destroyed everything that Nature took 200 million years to build. And as we said, we only have 2 minutes left relative to the time of the Universe, to avoid the complete destruction of Planet Earth.

And the inhuman humanity, will have to change definitively towards a better behavior, when as society, understands that all that political ambition is absurd, which even leads to the wars between brothers, simply for wanting to manage the economic resources that only belong to the Earth. Because misdirected knowledge, or in this way, is not used as an opportunity to guide those who are confused as true shepherds. And this irrational behavior exists only on Earth.

But finally, we understand that this is only done by those individuals who are in that process, or who comprise 35% of those who still do not deserve the title of being human beings, but humanoids, because they act only by instinct, and sometimes worse, than they themselves qualify as animals.

5

FORGIVE ME, EINSTEIN

The Einstein equation can actually be written as $E=(m-m_0)C^2=\Delta mC^2$ where m is the mass the particle acquires, only during its motion at the speed of light C; and, m_0 is the mass of the particle when it is still; or without motion.

But perhaps, as we said, the most significant or transcendental thing about this brilliant deduction, which arose from the mind of Albert Einstein, is that this equation could be tested experimentally, for those particles whose dimensions are on a subatomic scale.

And what we have done is to extend Albert Einstein's reasoning to the tiniest particles that exist, in order to imagine how the mass m arose from the rest mass m_0, when the Universe did not yet exist. And from there arose the mass that we can see, because what has not been condensed will be difficult to detect, or impossible to see with physical eyes or from this three-dimensional perspective.

And as shown in Figure 1, Bucherer and Neumann were able to prove in 1914 how the mass of an electron increases as its velocity increases with respect to an observer. And this was undoubtedly an event that revolutionized physics, because with this experiment, it was possible to prove that mass arose from the movement of the same particle. And so the law of relativity and Albert Einstein's mass were unequivocally and definitively established. The curved line is a graph of the square root of Einstein's mass m: $m = m_0\sqrt{1-v^2/c^2}$. (square root √) And the circles of experimental values have been adapted from the data of Bucherer and Neumann.

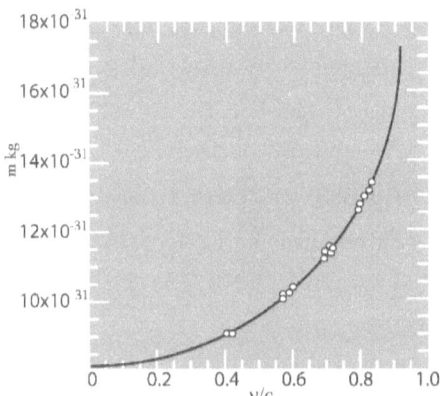

A graph showing the growth of the mass of an electron, as his speed increases.
Figure 1

But perhaps, what Einstein could not see, is that in reality, the curve tends to the infinite value, when the velocity U/C of the particle tends to the velocity C, that is, when U/C tends to the value 1 (U/C → 1) as can be seen in Figure 1. And from there we say: Forgive me Einstein, because in reality, and truly, there are particles that can move faster than light; or U/C>1. Which is significant, because when the particle velocity is larger, the acquired mass will also be greater, or $UE/C^3 = m_0$. And since C^3 is a constant, we can call it $\psi = C^3$, which means that $m_0 = UE/\psi$, in such a way that the mass m appeared by a proportionality of the velocity υ of the particle, with its energy E. And this, we can say, in honor of Albert Einstein, which is the mass m of Einstein. But that then Einstein's mass m would remain agglutinated, when other energies with sufficient integrating force act on it.

And in a general way it has been proven that for all kinds of energy, unlike potential or resting energy, these energies appear only by the action of some movement. For example the work ω, is the result of applying a force on a body to move a distance d. $\omega = F \cdot d$. In such a way that the energy in the form of work ω, will appear only when the force F is being applied on the body. Or perhaps the other example is when we compress a spring and give it a potential elastic energy U, then the mass of the spring, will increase from m_0 to $m_0 + U/C^2$, or when we add a quantity of heat Q to any object or system, the mass will increase in a quantity Δm; being $\Delta m = Q/C^2$.

And so we arrive at the principle of equivalence between mass and energy, which establishes that: for each unit of energy E of any kind, provided to a material object, the mass of the object will increase by an amount given by $\Delta m = E/C^2$. And this is Albert Einstein's famous equation; that is, the $E = \Delta m C^2$ equation that revolutionized, and to a great extent clarified, a good part of the great enigmas of the Universe. But we continue with this process, of how they were formed, or from where the Universe and the spirits really arose.

And in this way, all the rest mass or m_0 of the Universe could be created. Because when υ/C^3, or υ/ψ was made equal to 1 (one), the energy E of the almatrino became equal to the rest mass m_0, ($E = m_0$) so that energy is also called the rest energy E_0. $E_0 = m_0$.

And Albert Einstein wrote:

"Physics before relativistic theory, contains two conservation laws, which have great importance: namely, the law of conservation of energy, and the law of conservation of mass. And those two laws appear there, as complements to each other. But with the theory of relativity, both laws merge into one principle".

Of course, Einstein here refers to the fact that in classical physics mass is a constant when a body is accelerated; that is, Newton's mass; whereas in the theory of relativity, mass is relative to the speed of light and to the real mass of rest.

And so, we have reached the point where a single word would be needed, but which would be able to express with maximum emotion and exaltation all that can be said and felt; but this is limited to us by the way in which each one can write it to express it. But it was in this way that, in addition to the emergence of the Universe from almatrines and urdires, and from these the conscious energy itself; that is to say, the spirits, the great data was also formed, for the safeguarding of the most immense information of genetic succession that can exist in all the visible Universe. For these sequential units form the determined, defined, characterized and infinite configurations, so that from these encrypted data or the yotta tonalities formed by the almatrinos with their urdires, formed the 1×10^{24} units of consciousness. But from this condensation of energy, DNA was formed. but many believe that first RNA was like a ribozyme, that is, an RNA that can replicate by itself, and from there arose the infinite configurations of cells, which in turn, were organized to form each of the living beings: Let's say again, from a unicellular being here, and another one there, another microscopic, a plant, and where half of the information will be engraved by a single seed, or a spectacular salamander, because in it its spirit in the form of energy formed by almatrines, even manages to replicate itself as if it were a ribozyme, or to regenerate its amputated limbs. Or the cautious fish deep in the ocean, who looks very cautious from his cave that serves as a refuge, because there are many predators to the stalking ... It is a wonderful world. And this power of creation reached a human being, who emerges only when the almatrines and urdires take as their residence the mass that became a body.

And the point for a new beginning will depend on the sequence in which this information has been stored in the memory of the spirit by the almatrinos. But each time there is a new opportunity, there will be an update in those infinite configurations and possibilities, which only offers that data accumulated in quantum form.

But the fact that energy and matter were conserved at those molecular levels, as observed by Antoine Laurent Lavoisier, as a work of scientific reasoning, is something that failed to enter Albert Einstein's analytical reasoning. Inasmuch as, for Einstein, matter and energy were conserved only at the level of atoms and molecules. So Einstein had many doubts, that the same would happen for those subatomic particles. In such a way, that Einstein would imagine, that when he tried to project on the screen of his mind the movement of particles at a subatomic level, in this case, mass really became energy and energy again became mass, which is different from Lavoisier's case, when mass is transformed into another kind of mass, and energy into another form of energy. Because in Lavoisier's case, mass is not in motion. And at the moment or because of its great velocity, the mass of the particles must necessarily become what it was, that is, energy.

But it also turns out that Albert Einstein only managed to relate these movements relatively to the speed of light, because this luminous effect is the trace left by what Einstein called energy packages or photons, and it is the only thing that can be seen as bursts. But in addition, it is the maximum that can be measured in an equally relative way. Now let's hope that ψ is an absolute velocity constant for an elementary particle.

Because absolutely, that the speed of a particle ℧, is equivalent to the speed of light elevated to the cube. That is, C^3, or, 300,000 km/sec raised to the cube. Or the absolute speed of a particle ψ=27.000.000.000.000.000 km/sec And this is a value of a speed constant, truly enormous for the imagination of the human mind.

And it was certainly so, that indeed, with this definition of the mass of Albert Einstein, it was possible to analyze the kinetic behavior of these particles at the nuclear level. And in fact, that changed forever the concept that was rooted in scientific thought regarding the movement of particles, because it could be proved experimentally, that in reality, mass is created from energy. And for that, it is necessary that the particle is only in movement. Because if the particle is still, there will be no change in it.

And in the same way the Universe was formed, when the small space began to be disturbed. And at that moment, the almatrinos managed to accelerate until they managed to move with a speed of 27,000,000,000,000,000 kilometers per second. And they created all the existing energy in the Universe, as the only result of that immense movement of some very small particles; without load and without mass, because only the movement was required to create a minimum energy that later would be transformed into a minimum amount of mass. And there will only be growth of space, when there is some movement.

SPEED OF THE ALMATRINOS

And one of the most sensational events for science happened, because Albert Einstein's idea was demonstrated. That is to say, when a subatomic particle moves at a speed comparable to that of light, that particle acquires mass by itself. But this really has to be so, because as was said, moving mass acquires energy and this energy will be transformed into mass. But the experimental problem would be that at that time the particle accelerators were, so to speak, very rudimentary, and with the data obtained, it was only possible to mathematically extrapolate the progressive character of the phenomenon. Or because the force of acceleration achieved was not sufficient to reach at least the speed of light. But in addition, if we compare them, the particles used for the experiment, these were much bigger than the almatrines. Although the most modern detectors, or however sensitive they may be, will not be able to register these particles either, because the almatrinos will pass through them without leaving any trace.

In such a way, that we could not pretend to observe the phenomenon in a clearer and more extensive way, in order to go beyond Ʊ/C>1; that is, when Ʊ is greater than C, or C is less than Ʊ, because it was impossible to use a particle that could move, at least at a faster speed than light; that is, C. But it was even less thinkable, to be able to imagine the speed of an almatrino, nor was it suspected that the almatrino existed. So Albert Einstein, could not, or did not want to see beyond extrapolation, and only limited himself to analyzing the phenomenon, when Ʊ/C<1, because he concluded that nothing could travel at a speed greater than light. If this were the case, a mass m that could be moved the same as C would be converted instantly into energy. Or perhaps, because light is the only thing we can observe with the physical eye, although in a relative way. And it seems that nobody wanted or wants to see when Ʊ/C>1 because this actually violates Albert Einstein's theory of relativity.

But it is good to mention, that the accelerating capacity of these devices depend on the radius; that is to say, on the diameter of their physical design. And the speed of the particles does not depend on the frequency of the energy, but the faster particles move in larger circles, and the slower ones in smaller circles. For example, the accelerator under the Geneva mountains has a circumferential length of 27 kilometers. However, the hope that you can reach speeds higher than those of light, do not fade in me, because Chinese scientists will begin in 2030 the construction of the CEPC (Circular Electron Positron Collider), a particle accelerator that will have a circumference length of 100 kilometers in its path. And hopefully this concept of almatrinos can reach the hands of some Chinese scientist, so that at least try to perform in this new accelerator, particle collisions, at higher speeds than have been achieved so far in the accelerator in Geneva. Or in order to search for these particles, which are the smallest that exist.

And the equation $U=m_0C^3/E$, is the simplest thing that could have been deduced, to explain an extremely complex phenomenon. But this apparently simple equation explains how the Universe was formed; and then the spirits. But it was deduced from another very simple equation that Albert Einstein deduced; that is, $E=mC^2$. And the formation of the Universe, and everything that exists within it, undoubtedly has a meaning and a logical and simple trajectory. And perhaps the complexity of the problem, we put it at the time of searching and placing in order such explanations, such as the phenomenon of being able to travel relatively towards future events. We say relatively, because these journeys are relative to a person, or for those particles that move at a slower speed than light.

And all we have to do is look for explanations, or know how it is that energy becomes matter, and then how matter progressively becomes other kinds of matter, and energy becomes energy. But it is always the same matter coming from the same energy, which is the only thing that is being created by this unstoppable activity of the Universe. Because the equation that formed the Universe will be expressed in a very simple way, like:

$$U=m_0\psi/E$$

Of course ψ, is a constant of proportionality, and almatrinos do not have mass, but neither can we say that it is zero, because if we consider that m_0 is zero, we would make the phenomenon disappear mathematically. In such a way that we need to say, that the mass tends to the value zero, but it cannot be exactly zero, and we can consider m_0 within another constant that we will call $\Omega = m_0\psi$. Or that the value of that mass is the smallest that can exist. In such a way that $E=\Omega/\upsilon$.

So: $E=\Omega/\upsilon$. And the energy E did not exist either, when the Universe had not yet been formed. Because the energy E, appeared only when a disturbance occurred; that is to say, when the movement took place. And when the particle began to accelerate, until it reached the value of C, υ was small, and E became very large or tended to infinite value ($E\to\infty$). And thus was formed the great energy that managed to take the future Great Universe out of its stillness. And whenever the energy E, in the form of heat Q, appears, there will be a disturbance again, and what thus began can no longer be stopped.

And that center or dead center from which the Universe was formed, must still exist in a real way, but not in an imaginary way. And it has to be this way effectively, because the point from which the Universe was formed, it is impossible for it to disappear.

That is to say, the speed "υ" of the almatrino would be inversely proportional to the amount of energy E (the energy of the almatrino arising from movement) and the constant of proportionality would be the mass of the almatrino (m_a), in movement. Because when the particle is very tiny as in the case of an almatrino, and in this case smaller than the mass of a neutrino, the speed with which it moves will be greater, when the almatrino's energy at rest is less; that is, $\upsilon = m_a\psi/E$.

And it was in this way that an almatrino, being the most elementary particle, when it emanated a "quanta" of energy, managed to move in that small space, and managed to accelerate generating again the enormous energy. Enormous, relatively or with respect to that small space. Because if the energy was very low, and the almatrinos did not have mass, we can assure that in a beginning, or in that place, nothing existed as mass; and of this low speed, a great energy was formed, that for that small space was very intense, and thus a form of heat was generated in that small space, which was what managed to violently awaken the great Universe.

And for an almatrino, the mass gained, or m, will always be less than the mass of a neutrino. Being the neutrino, one of the smallest particles that science knows so far. Because the mass of a neutrino has been detected with enormous difficulties, for which reason, it is correct to think that the mass of the almatrinos, is not going to be able to be detected by any physical means that can be conceived in the human mind.

And the almatrines were integrated by the energetic force of the urdires. For which, we can rather deduce, that when the set of almatrino with their urdires, manage to diminish to will their speed, they become slow enough, and manage to show themselves as true energetic entities. And we will be able to see them from our three-dimensional perspective. But instead of spirits, we catalogue them as ghosts. But even so, the resting mass, or m_0 of the spirits, will indeed be too small; or null to say. For this reason, the spirit can cross any obstacle without being stopped. They can even pass through the gaps between the nuclei of the atoms of ordinary matter, as Ernest Rutherford observed in his experiment.

And that is what makes the spirits formed between the almatrines and the energy in the form of the urdires, as the force that acts in an integrative way, they cannot be seen by the naked eye or detected. Or photograph them. Unless they are constrained at will, and with it slow down their speed of movement to become visible before the lens of a camera. But these electronic devices have not yet managed to reach the resolution that the human eye has, so that the spirits can be detected as energetic apparitions, because the lenses of the cameras are pierced by the almatrines. Or the other way of manifesting, although they will remain invisible, is that the spirits incarnate occupying a body, taking an infinity of forms, like any living terrestrial Being.

And it is then, or when this process of being in a body ends, that the spirits may show themselves before us as ghosts or apparitions, because they have energetically copied the figure of their last physical form. Or as if they were energetic holograms of a high definition, because the integrating force of the urdires is very intense.

But this process is also relative, because we do not know if for them, that is to say, for those we call ghosts, the true ghosts are us. Because after understanding this process, it will really become a normal event for us, and we don't know if the day will come when the process of passing from one state to another will be done in a normal or daily way. The problem is that the cells can't stay long without breathing.

7

EQUATION THAT FORMED THE UNIVERSE

But now, we are going to prove mathematically, that almatrinos can actually move at a speed equivalent to the cube of the speed of light, i.e. 2.7×10^{16} kilometers per second (C^3). And this, perhaps the most connoted physicists will not be able to understand, but those of us who know, that we can in the astral state, move instantaneously from one place to another. Or practically without realizing it. Or that we can pass through any surface without any force being opposed to it. The light of photons, for example, is intercepted by an iron door, or even a sheet of paper, and to the light of spirits made by almatrines these surfaces seem not to exist, because nothing stops us in our trajectory as spirits.

And as we said, Albert Einstein deduced that when a particle moves at the speed of light, its mass effectively has to be transformed into energy, or that if this particle slows down, the same energy has to be converted into mass, and that is really so as we have already demonstrated. So, Einstein writes, that the mass created or gained by the moving particle would be given by equation (E-1). That is to say:

$$m = \frac{m_0}{\sqrt{1 - V^2/C^2}}$$

But if we apply this concept to almatrinos, because in the end we are analyzing them as particles, or more specifically as points, if we call $\mathsf{U} = R^2$ and $K = C^2$ and replace them in the Einstein equation we have what remains:

$$m = \frac{m_0}{\sqrt{1 - R/K}}$$

But we have given as a fact, that in reality almatrinos can travel at a faster speed than light, therefore:

If K<R implies that R/K>1 So

$$m = \frac{m_0}{\sqrt{-R/K}}$$

The value R/K of the root is a negative term, so we must multiply by (-1) and resort to the complex number i:

$$m = \frac{m_0}{\sqrt{-R/K(-1)}}$$

But K= C² and R= V², also √(-1) = i (the complex number)

$$m = \frac{m_0}{\sqrt{R/K}\sqrt{(-1)}}$$

$$\frac{m_0}{\sqrt{R/C^2} \cdot i} = \frac{m_0 C}{\sqrt{V^2} \cdot i} = \frac{m_0 C}{V \cdot i}$$

Substituting this value of m into the Einstein equation:

$$E = \frac{m_0 C C^2}{V \cdot i} = \frac{m_0 C^3}{V \cdot i} \implies V \cdot i = \frac{m_0 C^3}{E}$$

To eliminate the imaginary number, we will multiply by i, both sides of the equality:

$$V \cdot i \cdot i = \frac{m_0 C^3}{E} \cdot i \implies V \cdot i^2 = \frac{m_0 C^3}{E} \cdot i$$

$$V(-1) = \frac{m_0 C^3}{E} \cdot i \implies -V = \frac{m_0 C^3}{E} \cdot i$$

And to eliminate i, we raise the modules squared on both sides:

$$\left|-V\right|^2 = \left|\frac{m_0 C^3}{E} \cdot i\right|^2 \implies -V^2 = \frac{m_0^2 C^6}{E^2} \cdot i^2$$

Once again $i^2 = -1$

$$-V^2 = \frac{m_0^2 C^6}{E^2}(-1) \implies V^2 = \frac{m_0^2 C^6}{E^2}$$

$$V = \sqrt{\frac{m_0^2 C^6}{E^2}} = \frac{m_0 C^3}{E}$$

In such a way, that the square root of v, or the speed of the almatrino is:

$$V = \frac{m_0 C^3}{E}$$

(E-2)

This equation would represent, by definition and deduction, the speed of an almatrino, or what is known as a tachyon. But as we have said, the speed is variable. But in addition, this velocity is not really inherent in itself, like for example the velocity of a photon. Which means, that the movement is proper or characteristic of the almatrino. That is why we prefer to call it Ʊ, as a way of trying to better define, instead of a speed parameter, the maximum speed at which an almatrino can move.

So that the mass ma is really the mass of the almatrino when it is stopped and E_a is like saying the potential energy contained in the almatrino when it is immobile; so:

$$V = \frac{m_0 C^3}{E}$$

(E-3)

And the equation (E-3) is the most important that has been deduced, because it explains how the Universe was formed. But it will also help us to clarify a large number of other questions. For example, the ones I could observe from when I was a child. Because with this equation, now I can explain why I could see ghosts; or why I could get out of the body; or why I could go through doors, and when I got out of the body, I could get out of the room and observe the outside world. But not as a dream, but in a real way. Or why I could see future events. In addition to other concerns. But these we cannot consider here, because they have a strong involvement in the religious field. Although all these, and the other conclusions will remain free, so that others will analyze them.

And with the deduction of the equation $U = m_a C^3 / E_a$ what we are concluding, that in the beginning, in reality the Universe did not really have any energy, because the energy appeared only when the speed U of the almatrinos really tends towards a value that is infinite.

And we can conclude, that in the beginning there was nothing. No mass, no energy. For there was only the smallest dead center that we can imagine, occupied only by the smallest particle that can exist, and which we have called almatrino. And we call it dead point, only as a way of defining it, because in that particle there were no forces of vibration or rotation, for example. But when this particle was able to move, it was from this movement that the energy arose that exploded the small energy bubble that awakened the creation of the Universe. And from where all that exists originated, which is only energy. Because matter is nothing but the same energy that arose, but was agglutinated by another kind of forces, equally energetic that integrated them, and these integrating forces is what we call bosons. But the real thing is that it is the same energy emanated. And if we manage to collect the Universe again, to bring it to that dead point, we would have to cool it at the same time, because we would not be able to concentrate, in that point, all the energy that has already been generated. Because, in addition, in the beginning, in that point the energy did not exist, which is opposite to what is raised with the mass and the energy of Max Planck.

ABOUT THE AUTHOR

Graduated from the School of Chemistry, Faculty of Sciences of the Central University of Venezuela, with a degree in Chemical Technology. Postgraduate studies in Food Science and Technology. Special work on the chemistry of natural products and the chemistry of diseases. Study of cosmology and the origin of spiritual energy.

www.ingramcontent.com/pod-product-compliance
Lightning Source LLC
Chambersburg PA
CBHW021923170526
45157CB00005B/2155